Tractors

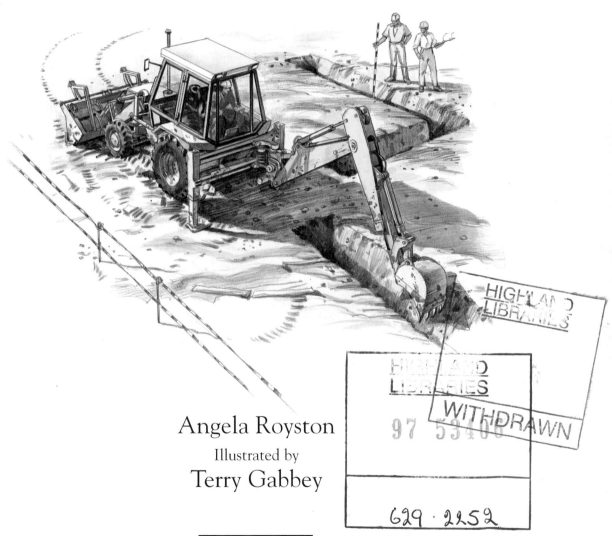

Angela Royston

Illustrated by
Terry Gabbey

Contents

Tractors 4

Early Tractors 6

Hay-making 8

Driving a Tractor 10

Ploughing 12

First published in Great Britain in 1998 by Heinemann Library,
Halley Court, Jordan Hill, Oxford, OX2 8EJ,
a division of Reed Educational & Professional Publishing Ltd.
Heinemann is a registered trademark of Reed Educational & Professional Publishing Ltd.

OXFORD FLORENCE PRAGUE MADRID ATHENS
MELBOURNE AUCKLAND KUALA LUMPUR SINGAPORE TOKYO
IBADAN NAIROBI KAMPALA JOHANNESBURG GABORONE
PORTSMOUTH NH CHICAGO MEXICO CITY SAO PAULO

Editor: Alyson Jones Designer: Nick Avery
Picture Researcher: Liz Eddison Art Director: Cathy Tincknell
Printed and bound in Italy. See-through pages printed by SMIC, France.

02 01 00 99 98
10 9 8 7 6 5 4 3 2 1

ISBN 0 431 06549 7

British Library Cataloguing in Publication Data
Royston, Angela, First look through tractors
1. Tractors - Juvenile literature 2. Agricultural machinery - Juvenile literature
I. Title II. Tractors 629.2'252

Acknowledgements
The Publishers would like to thank the following for permission to reproduce photographs: page 5: Art Directors Photo Library © Trip/D Houghton;
page 7: Tony Stone Worldwide © Bruce Hands; page 8: Tony Stone Images © Peter Dean; page 15: © Britstock-IFA/Frisch;
page 17: © Christine Osborne Pictures; page 19 top right: © ZEFA-CPA; page 19 bottom right: ZEFA © Smith R.;
page 21: Tony Stone Images © Mitch Kezar; page 23: © Bomford Turner.

Any words appearing in the text in bold, **like this**, are explained in the Glossary.

Sowing Seeds 14

Spraying Crops 16

Harvesting Crops 18

Harvesters 20

All Kinds of Jobs 22

Index and Glossary 24

Tractors

Farmers use tractors all year round. Tractors pull trailers and farm machines, such as ploughs, balers and crop-sprayers. They have big wheels with thick tyres and can be driven over stony ground and through deep mud.

This crawler tractor has tracks instead of wheels. It is used where the ground is full of sharp, flinty stones. These stones would rip up rubber tyres.

Not all tractors are used on farms. This tractor is towing a boat out of the sea. The boat is on a trailer.

Early Tractors

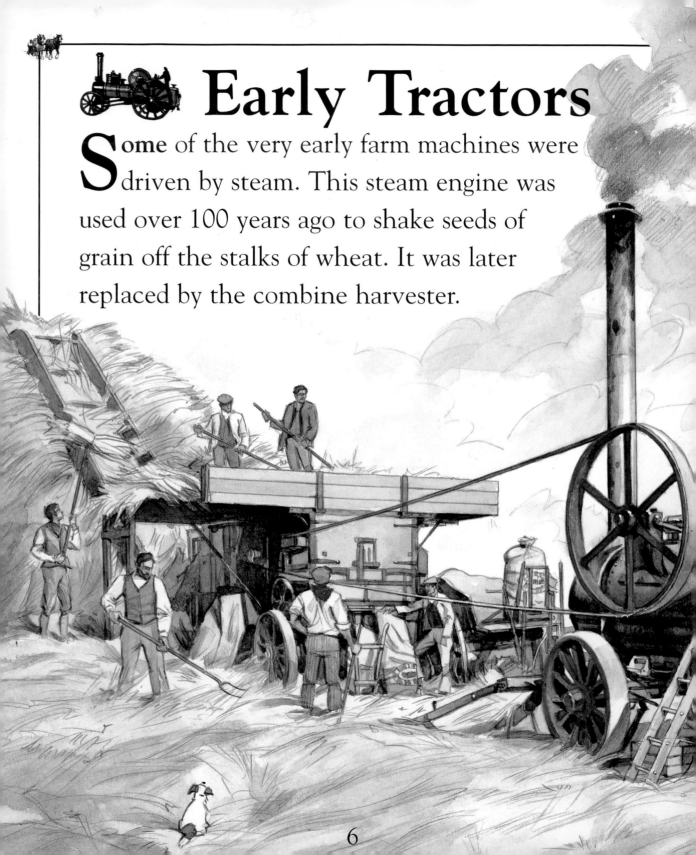

Some of the very early farm machines were driven by steam. This steam engine was used over 100 years ago to shake seeds of grain off the stalks of wheat. It was later replaced by the combine harvester.

This early tractor was designed in 1908 by an American called Henry Ford. He called tractors '**automobile** ploughs'. Ford also **invented** the first cheap car.

Before tractors were invented, horses and other animals pulled farm machines. Look how many horses this **Amish** boy is using to pull his **plough**.

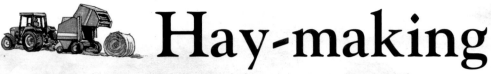

Hay-making

Grass grows quickly in spring. When it is ready to be cut, the tractor pulls the mowing machine up and down the field. The cut grass is left to dry and becomes hay.

Some farmers do not wait for the grass to dry into hay. This machine chops the freshly cut grass and blows it into a trailer. The wet grass is stored as silage.

Silage is stored and fed to cows and other animals in the winter when there is little grass for them to eat.

The farmer then brings a baling machine to the hay-field. It gathers and presses the hay into big bales. What is the other tractor doing?

Driving a Tractor

Inside the cab of a tractor there are lots of levers and switches. The farmer moves these to work whatever machine the tractor is pulling. In winter a heater keeps the cab warm, and in summer the roof lifts up to let in fresh air.

This tractor is pulling out a car that was stuck in the snow. A light flashes on top of the cab to warn other drivers that the tractor is there.

The driver of this tractor has fixed a special blade to the front of it. The blade pushes the snow to the side of the road. Now the road is clear for cars.

Ploughing

Before they can plant seeds, farmers prepare the land. The soil needs to be broken up to make it easier for the seeds to grow. This tractor pulls a plough over the field. The blades dig deep into the soil and turn it over.

Watch out! This farmer is spreading **manure** and straw over the land. The manure feeds the soil and makes the seeds grow better.

The round wheels of the harrow behind this tractor break up the soil into smaller pieces. Now the farmer can sow the seeds.

Sowing Seeds

Seeds are carried in a flat box above the drill. The sharp spikes of the **drill** cut narrow trenches in the soil. As the seeds drop into the trenches, the machine covers them with soil.

These three women are planting tiny lettuces. They push them into the soil as the tractor moves slowly across the field.

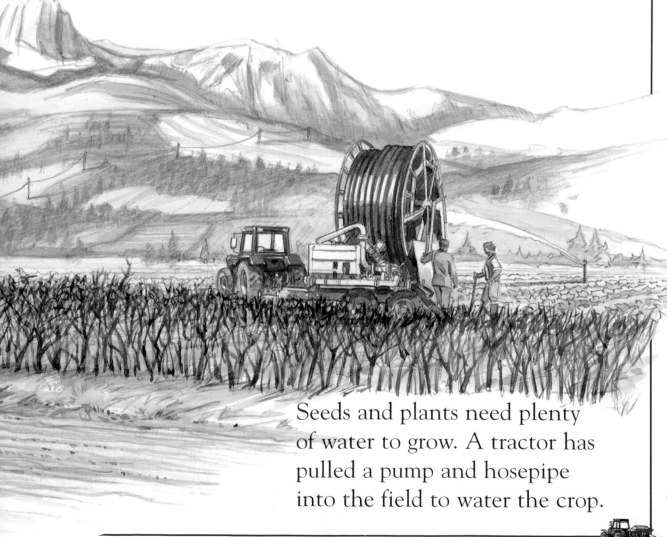

Seeds and plants need plenty of water to grow. A tractor has pulled a pump and hosepipe into the field to water the crop.

Spraying Crops

Sometimes farmers spray their crops with chemicals to kill insects or diseases which attack the plants. The sprayer is hooked onto a tractor and the chemicals are poured into the sprayer's tank.

Look at the huge tank at the back of this tractor in Malaysia. It is spraying the cocoa trees with chemicals to kill pests and diseases that are attacking the trees.

The farmers unfold the spray arms. As they drive through the field, a fine spray of chemicals covers the crop.

Harvesting Crops

Special tractors **harvest** different kinds of crops.
This machine chops the tall canes of sugar and
strips off their leaves. The tractor takes the
canes to the sugar mill where the sugar
is squeezed out of them.

What would you rather do – pick beans by hand or drive this amazing bean-picking machine? The blades at the front scoop up the beans and they are fed into the trailer at the side.

Look at this huge field of cotton. This tractor picks the fluffy white balls of cotton off the plants and stores them in the trailer behind.

Harvesters

This huge **prairie** is covered with ripe, yellow wheat. The combine harvester works fast. A spiky blade cuts through the stalks of wheat and a **conveyor belt** feeds the wheat into the machine.

This field of barley is so big, several combine harvesters work together to cut all the grain. The cut stalks will be made into bales of straw.

Inside the combine harvester, the little grains of wheat are knocked off the stalks and stored in the tank.

When the tank is full, a tractor and trailer drives alongside the combine harvester. Can you see how the grains get from the tank to the trailer?

All Kinds of Jobs

A **backhoe** loader is a special kind of tractor. The driver can face backwards or forwards. The loader at the front lifts heaps of straw and manure. The **hoe** at the back is good for digging holes. Other tools can be fitted to do special jobs.